T0060785

The Boy Who Dreamed of Infinity

A Tale of the Genius Ramanujan

The Boy Who Dreamed of Infinity

A Tale of the Genius Ramanujan

Amy Alznauer

illustrated by Daniel Miyares

Candlewick Press

First edition 2020

Library of Congress Catalog Card Number 2020902246
ISBN 978-0-7636-9048-9

22 23 24 25 CCP 10 9 8 7 6 5

Printed in Shenzhen, Guangdong, China

This book was typeset in Dolly.
The illustrations were done in ink.

Candlewick Press
99 Dover Street
Somerville, Massachusetts 02144

visit us at www.candlewick.com

To my parents with infinite love
A. A.

For Stella and Sam, may you always
dream beyond the stars
D. M.

Today the world is small. With a single click, you can see anywhere or speak to anyone.

But one hundred years ago, the world was big. It took weeks to send a letter by steamer from here all the way to there.

Back then if you had an idea — even a rare and wonderful idea — on one side of the world, people on the other side might never know. Your ideas might forever stay where they began, right here . . .

. . . in a small village in South India, by the banks of the Kaveri River, in Amma's arms.

Amma held her new baby. She touched his tiny pink tongue and told him about his grandmother's dream.

"Before you were even born," she said, "the goddess Namagiri leaned down and whispered in your grandmother's ear that someday she would write the thoughts of God on your tongue."

Amma gazed into her baby's bright eyes. "But how could someone so small ever grow up to say something so big?" Through the window with no glass, she showed him the stars.

"I will call you Ramanujan," she said.

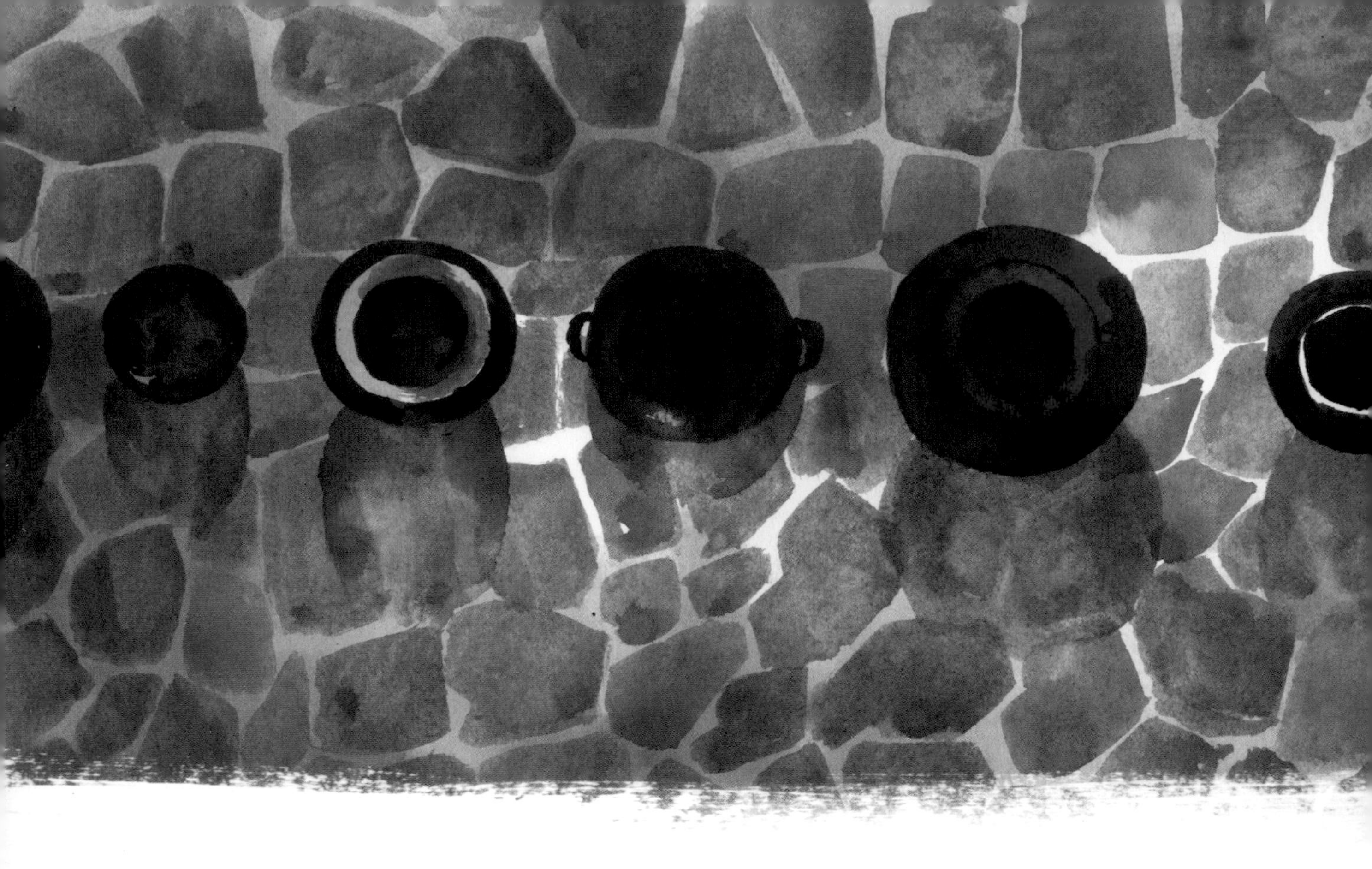

For three years, he stayed quiet as a mouse.

"*Chinnaswami*, my little lord," Amma begged, "what are you thinking?"

Ramanujan just lined up the copper pots across the floor.

And when he didn't get his curd rice and mango, he rolled in the monsoon mud.

His grandfather tried to help. He held Ramanujan's finger and together they traced numbers in rice. "*Onru, irandu, moonru,*" his grandfather counted out loud in Tamil. One, two, three. Soon Ramanujan began to talk.

"What is small?" he asked. He imagined the world with only one man, before anyone was there to hear him speak.

"And what is big?" He looked up at the big blue space between clouds.

At five years old, long hair up in a knot, white *dhoti* tied around his waist, he started school. Just half a dozen boys on the front porch of a house. But when the teacher took no interest in his odd questions, Ramanujan grew bored. He tried again and again to sneak away.

As he grew, he often thought about his grandmother's dream. There was something out there, beyond his tiny stucco house, whispering. Was it calling him? *Naa, naa, naa,* said the goats in the street.

What else is small? Ramanujan wondered. He remembered the legend of the single egg that cracked open to reveal the entire universe. He thought about a mango.

A mango is like an egg. It is just one thing. But if I chop it in two, then chop the half in two, and keep on chopping, I get more and more bits, on and on, endlessly, to an infinity I could never ever reach. Yet when I put them back together, I still have just one mango.

He loved this idea, small and big, each inside the other. If he could crack the number 1 open and find infinity, what secrets would he discover inside other numbers? It felt like he was setting out on a grand chase.

Numbers were everywhere. In the squares of light pricking his thatched roof. In the gods dancing on the temple tower. In the clouds that formed and re-formed in the sky. Every day he wrote numbers in the sand, on his slate, on slips of paper, his slender fingers flying, each number a new catch.

In school, numbers were different, stiff and straight, obeying the rules. When the children chanted sums, Ramanujan thought, *But aren't there more sums inside of those sums?* He broke the numbers into smaller and smaller bits, until he had only 1s. He found all the sums they had missed.

Is every number like this — a single body, but with many parts? Ignoring the class, he scratched wildly on his slate, his elbow darting up to erase.

“What are you thinking,” said the schoolmaster, “making such a racket on your slate and refusing to listen?” He made Ramanujan put his finger—*shhhhhh*—over his lips and sit in the corner.

Ramanujan hated school. Once, the town constable had to chase him back to his studies. So every year his parents tried a different school. They finally settled on Kangayan Primary.

Amma sang songs in the temple. Appa sold fabric in a shop. His friends played *goli gundu*, shooting marbles in the street. But Ramanujan was shy and liked to be alone.

Every day, he rushed home from school to sit on the front *pial* of his house, long hair coming loose from its knot, slate across his knees. Often it looked like he was staring off into nowhere. But it was not nowhere. Out there, numbers leaped and roared. He could see them even when he closed his eyes.

"Are you a genius?" a friend called up to him one day.

"I don't know," Ramanujan called back, "but if I am, my elbow is making a genius of me!" He held up his elbow, smudged from erasing the slate. Then he was at it again, writing and erasing, as if perfecting a sketch.

What else is big? he needed to know.

If I can break a number into so many sums, how many sums do I get if I take bigger and bigger numbers? When will it be too big to count? He wrote until his fingers ached.

In the summers, when it was too hot to think, he chalked numbers on the cool temple floor. High up in the pillars, bats fluttered. He was like an archaeologist, dusting off the hidden bones of each number to find its structure.

Do numbers also break apart by multiplying? he wondered. Now he ended up with numbers like 2 and 3 and 5, with no smaller pairs inside them. *How many numbers like this are there? And how do I find them?* There was no book with an answer, no one to ask. He just had to write and think.

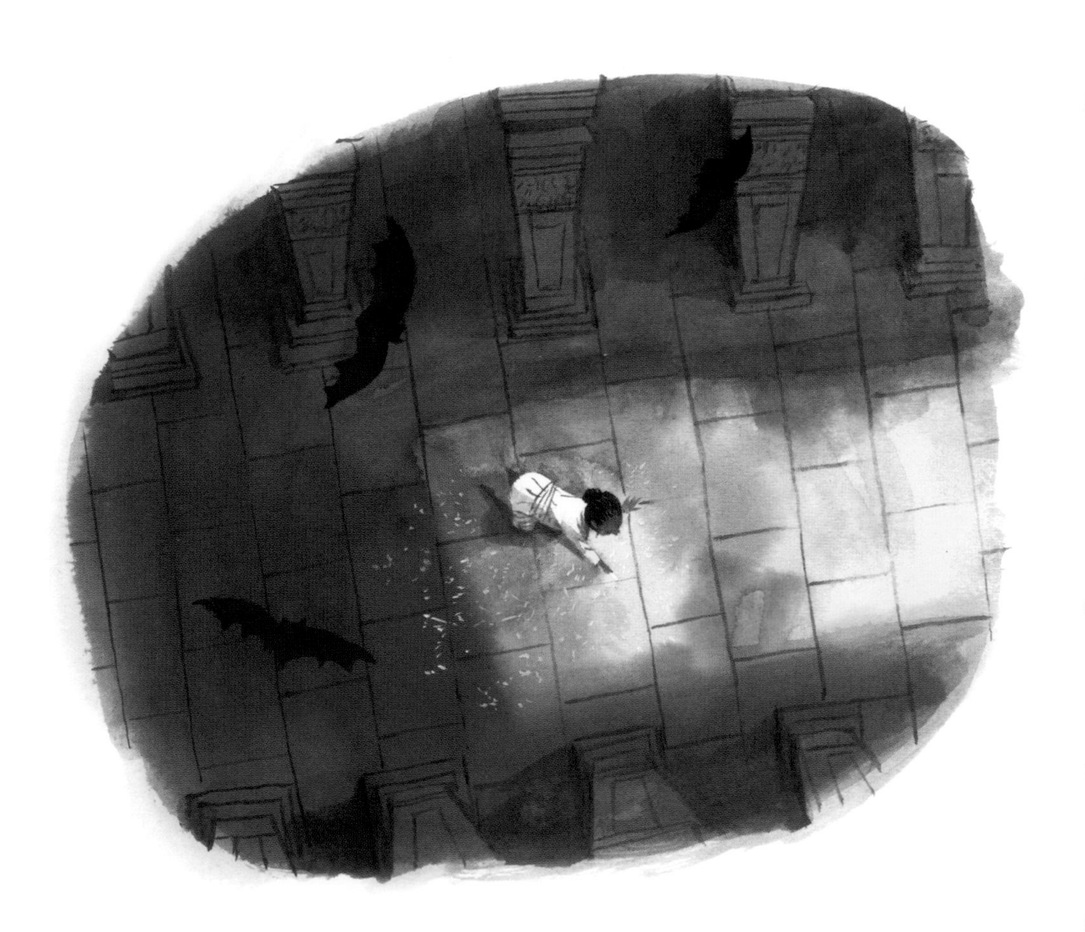

In the evenings, while Ramanujan scribbled away and stuffed sticky rice balls in his mouth, Amma sang about the gods. She told of the tiny fish who grew to be as big as an ocean and the little boy who once saw a great golden lion-man break out of a pillar of stone. *Everything in the whole world is about small and big*, thought Ramanujan. He watched a spark float up and away from the fire.

And at night, while he slept, numbers came whispering in dreams. When he woke, an idea would be there as if placed on the tip of his tongue. He'd write furiously, trying to catch the golden thought before it fled.

When Ramanujan turned ten years old, he entered Kumbakonam Town High School, where a schoolmaster finally saw that he was lightning-quick. He solved tricky problems in a flash. He made up magic squares to stump his friends.

Often Ramanujan was quiet, but when his classmates got him talking, the words gushed out and his eyes glowed, mischievous and sparkling. He was like no one they'd ever known.

He took food to the man who ranted down by the Kaveri River. The man stared into the sun and said he could see odd creatures running about. But this didn't seem strange at all to Ramanujan.

"Sometimes even invisible things can be real," he told his friends.

When he was fifteen years old, he got his hands on a college math book. It had thousands of questions but no answers. Ramanujan answered the questions. All of them. And when the questions ran out, he made up more.

He started writing down his ideas in a notebook with a green pen. The blank pages thrilled him. All that paper just waiting for his work. Ramanujan tried to write neatly, but then he'd catch the whisper of an idea. Numbers would rush across the pages in circles and packs.

Amma watched him, remembering the dream. *Are these the thoughts Namagiri placed on his tongue?* she wondered.

As Ramanujan grew older, his questions grew bigger.

How many sums are inside two hundred? How many numbers like 2 and 3 and 5 are there up to ten million? It would take an impossibly long time to count that high. He discovered intricate formulas and patterns to answer his questions.

No one told him how math is supposed to be done, so he did it his own way. He devised his own symbols. He worked out ideas no one had ever thought before.

Will an endless list of tiny numbers add up to 1 or infinity? Will it be like the bits of mango or the countless stars in the sky? He found beautiful ways to tell different lists apart.

But when Ramanujan entered college, he couldn't focus on anything except math, and he flunked out. At twenty years old, he had nothing. No college degree. No job. Only a notebook covered in strange green ink.

He was back on the *pial* of his house, writing fervently on his slate.

Sometimes his family didn't have enough to eat, so a neighbor fed him butter and rice. Amma put her foot down. She'd had enough of sitting on porches and writing on slates. She arranged his marriage and demanded that Ramanujan get a job.

Carrying his notebooks under his arm like a diploma, he went door to door asking for work. Everyone saw that he was brilliant, maybe even a genius, but they didn't quite know what to do with him. He finally landed a job as a clerk keeping accounts at the Port Trust in Madras.

Still, he did math every chance he got, sometimes late into the night, scraping on his slate. One night, tired of the noise, a friend dumped a bucket of water on his head and cried, "What are you thinking?"

Ramanujan looked up, his long hair dripping, his eyes shining. "I am small," he said. "God is big. I am trying to learn the thoughts of God."

He often walked on the long Madras beach and looked out at the ocean. *I am like the first man in the world with no one to hear me speak*, he thought. He watched the steamers pass in the distance. Were these thoughts really meant for him alone? Secrets to be caught and kept forever? Wasn't there anyone, anywhere out there, he could tell?

Local mathematicians and British overseers at the Port Trust became convinced that Ramanujan was extraordinary, but nobody they knew was working on similar mathematical ideas. Was anybody in the world? They urged him to write a letter.

"I beg to introduce myself," he began, and wrote out his strange symbols and ideas. He sent the letter by steamer to Cambridge University, one of the great mathematical centers in England. Weeks went by. Nothing. He tried again. And again.

Ramanujan would try one more time. He wrote to G. H. Hardy, one of the top mathematicians at Cambridge. He had recently discovered Hardy's pamphlet on infinity. Maybe Hardy would understand. Ramanujan waited . . . and waited.

Finally a letter came back and Ramanujan tore it open. Hardy was astonished. How did Ramanujan come up with such outlandish, magnificent ideas? And would he come to England?

Ramanujan's whole life had been building to this moment. He desperately wanted to share his work, but could he really leave his home?

He went to the temple to pray. After some time, a whisper seemed to come in a dream: "*Speak the thoughts on your tongue.*"

So, Ramanujan said goodbye to Amma and Appa and his young wife. He cut his long hair. And on a bright March morning in 1914, he packed up his notebooks, took one small step onto the steamer, and set out from here all the way to there.

As he rocked on the steamer and gazed up at the great night sky, so full of stars that it looked like a glittering infinity, he never could have guessed that someday scientists would use his ideas to help explore that sky and that his work would change the course of mathematics forever. One hundred years later, people would still search his notebooks in wonderment, trying to discover what he was thinking.

Acknowledgments

Thank you to Kumar Sambandan for his help with Tamil words and sounds; Krishna Alladi and G. P. Krishnamurthy for their generosity when I was doing research in India; Bruce Berndt and my father, George Andrews, for their mathematical guidance; Robert Kanigel for his literary advice and incomparable biography of Ramanujan; the Number Theory Foundation and the Indian Room at the Cathedral of Learning in Pittsburgh for research grants; Esther Hershenhorn and the SCBWI-Illinois Laura Crawford Memorial Mentorship for guidance; Rosemary Stimola for expert representation; Hilary Van Dusen for enthusiasm and clarity; and my wonderful family—for everything.

Author's Note

When I was six years old, I took a trip to England with my father, who is a mathematician. All I remember from that trip is my father falling into the river Cam, still holding the punting pole and splashing to the bank as the rest of us drifted away. Looking back, I know his thoughts were racing. He had just discovered Ramanujan's Lost Notebook, forgotten for decades in a box of papers that had found its way into Cambridge University's Wren Library. This discovery has been likened to someone unearthing the tenth symphony of Beethoven. And the romance of this story is one of the reasons I studied mathematics in graduate school and still teach it today.

It is the exuberant freedom and creativity of Ramanujan's quest that excites me. So in this book I focused on the development of Ramanujan's young mind and his struggle to be understood. As a boy, he set out into the wilderness of numbers with the zeal of an explorer. He had only inspiration and hard work to guide him.

During Srinivasa Iyengar Ramanujan's time in England, people came to see him as one of the greatest mathematicians the world has ever known and voted him into the Royal Society. He enjoyed a brilliant collaboration with G. H. Hardy, which Hardy said was "the one romantic incident" in his life. But, tragically, Ramanujan became fatally ill and, after returning to India, died at just thirty-two years old. It's incredible that he was able to pack several notebooks so full of amazing ideas that people are still working to understand them today. His ideas have helped shape areas of science not even discovered in his lifetime: computers, black holes, and string theory. But more than anything else, the profound originality of his ideas has been a source of inspiration for mathematicians ever since.

India has a rich mathematical history. Centuries before the Greeks, Indian thinkers developed sophisticated numbering systems, geometric identities, and both the concept and symbol for zero. And numbers weren't just for scientists but woven into everything: philosophy, art, religion. Ancient Vedic priests would rise in the morning and say, "Hail 100! Hail 1,000! Hail 10,000!" and on and on until they counted so high, all that was left to say was "Hail sun and all that rises!"

Every morning when Ramanujan woke, numbers rose with the sun and spread to every corner of his mind. From the time he was a little boy, he learned mathematics not from books or teachers but by experimenting with numbers himself. He dusted off their bones and looked for their hidden structures. So I show him discovering ideas like primes and partitions and infinite sums even before he knew what to call them.

Ramanujan was a number theorist, a person who studies the properties and patterns of numbers. J. E. Littlewood once said that "every positive integer was one of Ramanujan's personal friends." And Ramanujan shocked Hardy one day by observing that taxicab number 1729 was quite interesting. "It is the smallest number," he said, as if he'd always known, "expressible as a sum of two cubes in two different ways." (The two ways are $1^3 + 12^3$ and $9^3 + 10^3$.) This happens to be a direct quote, but many of Ramanujan's thoughts and words in this story are imagined, yet inspired by things he actually said, like "An equation for me has no meaning unless it expresses a thought of God."

What is so beautiful about this area of mathematics is that the greatest, most difficult questions can often be understood by young children. Maybe someday you, like Ramanujan, will catch the whisper of an idea and set off on your own mathematical quest.

Bibliography

Andrews, George E. "An Introduction to Ramanujan's 'Lost' Notebook." *American Mathematical Monthly* 86, no. 2 (February 1979): 89–108.

Berndt, Bruce C., and Robert A. Rankin, eds. *Ramanujan: Essays and Surveys*. Providence, RI: American Mathematical Society, 2001.

Dimmitt, Cornelia, and J. A. B. van Buitenen, eds. *Classic Hindu Mythology: A Reader in the Sanskrit Purnas*. Philadelphia: Temple University Press, 1978.

Hardy, G. H. *A Mathematician's Apology*. Cambridge: Cambridge University Press, 1940. Reprint with a foreword by C. P. Snow, 1967.

———. *Ramanujan: Twelve Lectures on Subjects Suggested by His Life and Work*. Providence, RI: American Mathematical Society, 1999. First published in 1940 by Cambridge University Press.

Kanigel, Robert. *The Man Who Knew Infinity: A Life of the Genius Ramanujan*. New York: Washington Square, 1991.

Letters from an Indian Clerk. Directed by Christopher Sykes. Independent Communications Associates, 1987.

Plofker, Kim. *Mathematics in India*. Princeton, NJ: Princeton University Press, 2009.

Srinivasan, P. K., ed. *Ramanujan Memorial Number*. Vol. 1 of *Letters and Reminiscences*. Madras: Muthialpet High School, 1968.